CHARLES DARWIN

A Great Biologist

THE HISTORY HOUR

HISTORY

CONTENTS

I

THE EVOLUTION OF DARWIN

"A man's friendships are one of the best measures of his worth."

— CHARLES DARWIN

HOW A MAN WITH NO MISSION BECAME ONE OF THE MOST DRIVEN IN HISTORY

Charles Darwin is one of the most famous names in history and for good reason – he developed the original theory of evolution, which speculated that humans and other animals evolved over time. This theory directly challenged the common way of thinking in the Western world, which was controlled by Christian beliefs. Namely, it directly challenged the religious idea that God made humans in His image and that He had created all the animals as they were by the 19th century. How did a man who would play such a pivotal role in the development of contemporary scientific thought begin?

Well, Charles Darwin led the normal life of a young man from a wealthy British family in the 19th century. Continuing the social, educational, and financial limitations of the period, is highly unlikely he would have ever developed his career without the backing of family status and the ability to attend university. Charles was the fifth of six children, born in 1809 and living until 1882. His family home was in Shrewsbury of Shropshire, and he was raised with the expectation that he would be like all his previous male relatives and become a professional in the medical field. However, his familial home did have some characteristics which made it special compared to others and helped Charles develop his interests – namely, close relatives who were also fascinated by natural science and philosophy.

❧

Like many great thinkers from the 19th century, Charles Darwin received a privileged start to life by being born into a family of wealthy intellectuals benefitting from the Industrial Revolution in England. Although Charles Darwin would not be born until 1809, it's important to set the scene for the kind of household which spawned a man willing to challenge preexisting theories and notions about the Christian universe. Darwin's intellectual inclinations were heavily influenced by the ideas of his grandfather, a man named Erasmus Darwin (1731-1802).

❧

This first Erasmus Darwin did not live to see his grandson's birth but laid the foundation of intellectualism and interest in natural science which characterized the Darwin family for centuries. He worked as a physician and as a poet on the side and developed an interest in natural philosophy young. His family background gave him a decent start in life, but his wealthy patients pushed him into the upper echelons of society. Amongst these patients was his greatest patron and dearest friend, the famous Josiah Wedgwood. Both men attained membership in the Lunar Society, an organization for wealthy men interested in new scientific and technological developments that only met during the full moon – they mockingly called themselves "***The Lunatics***" with airs of jovial familiarity.

Erasmus Darwin's interests, wealth, and ability to remain abreast with new developments in England allowed him to turn his attention to a subject dear to his heart – natural science and philosophy. The Lunar Society was already a progressive society that supported religious freedom, American dependence, and the end of slavery, so he found men willing to listen to his ideas about the evolution of living creatures. In 1794, towards the end of his life, Erasmus published a book entitled Zoonomia; or the Laws of Organic Life. Zoonomia was a two-volume medical work that made observations about plant cross-fertilization, the domestication of animals by humans, and the concept of inheritable characteristics. Many scholars view it as a significant work which incorporated early ideas for Charles Darwin's theory of evolution years later.

Erasmus Darwin's progressiveness and willingness to challenge existing intellectual ways of thought passed down to his son, Dr. Robert Waring Darwin, or Charles' father. Charles Darwin looked up to his father and always referred to his as the greatest or kindest man when speaking of him. Dr. Darwin married Susannah Wedgwood, the daughter of Josiah. She also brought into the family emphasis on scientific, technological, and intellectual development which heavily shaped and influenced the young Charles' education. Unfortunately, she died in 1817 when Charles was only eight years old.

Charles' family pedigree prepared him for a life of curiosity and interest in the natural world but did not make him a good student, especially his mother's death made his father far more stern and overbearing than he had been previously. Charles started to attend a boarding school in 1817 at a location in Shrewsbury after his mother's death. Since the school was only a mile from his family home, he frequently returned to the house on weekends and spent his time collecting beetles and performing chemistry experiments with his brother in a garden shed.

He did not do well in this first school. From a young age, Charles was more interested in natural and scientific subjects, and the school he attended was focused on classical studies. This meant he learned Latin, Greek, a little bit of geography, and a little bit of ancient history. He neglected his studies for years and eventually fell out of love with chemistry around the age of 15, choosing instead to spend most of his time hunting. This angered Dr. Darwin, who chastised his son and accused him of heading down a path where the young Charles would never amount to anything. Charles was pulled from the school and started as a medical assistant as his father's practice, where he would take notes and attempt to make diagnoses about patients – it can be safely assumed that he most likely did not actually diagnosis the patients.

❧

Charles' time as an assistant prepared him to go to the same medical college as his father and older brother, also named Erasmus (nicknamed Ras). He started at Edinburgh University in 1825 at the tender age of 16. Charles was bored. He hated the medical lectures and couldn't stomach the surgical theater, where professionals cut people open without anesthetics. He and Ras instead spent much of their time being wealthy young men in a large city, which meant a lot of playing and drinking.

❧

It was at Edinburgh University that Charles did discover his passion, though, and that was the natural world. He enjoyed visiting local fisherman and looking at the animals and organisms caught in their nets and orbited around Robert Edmond Grant, a physician, and zoologist who specialized in the study of sponges. He seemed to stimulate Charles' interest in marine invertebrates. Charles' other emergent passion was taxidermy of all things, inspired by his attending of a lecture by John James Audubon, an ornithologist, and painter from the United States. Charles learned taxidermy from an intriguing former slave named John Edmonston, who had traveled to South America and frequently told stories of adventures. The two became fast friends and kept in correspondence for years.

❧

During this second year of schooling, Charles visited Paris to meet with his family, including his cousin and future wife, Emma Wedgwood. During this time in 1827, his three older sisters begged their father not to force Charles to attend medical school any longer, seeing as the young man simply had no interest in the field. When Charles returned to England, Dr. Darwin relented but insisted that his son should now become a clergyman. Charles did not wholeheartedly agree with the religious ideas of the Church of England but being a country parish appealed to his sensibilities. He rationalized this career change meant he would be able to spend more time on his nature studies than if he had become a doctor, an idea which had some weight to it.

❧

Charles found himself sent to Christ's College, Cambridge after spending roughly eight months studying and brushing up on his Latin and Greek, since his language skills were not good enough. During this time, he developed a relationship with a young woman named Fanny in Shrewsbury, the friend of one of his sister's. It continued for a long time until he did not come home for Christmas. Fanny was offended and decided to pursue another suitor.

❧

The environment at Christ's College suited Charles much better. He was registered in the ordinary Bachelor of Arts program and spent most of his time riding, hunting, and participating in the growing fad of beetle collecting. Charles was so good at finding beetles that he managed to have some of his work published in James Francis Stephens' Illustrations of British Entomology.

❧

When not drinking and partying, Charles managed to become close friends with his botany professor, John Stevens Henslow. The Reverend Professor Henslow was perhaps Charles' greatest friend in life, and Charles himself said:

"I fully believe a better man never walked this earth"

in a letter to another friend in 1861. Henslow actively encouraged Charles' interests in the natural world. He brought the student on field trips, had active learning sessions where Charles and the other students could gather plant and animal samples, and frequently invited Charles over for dinner at his home.

Henslow opened the world for Charles Darwin. He offered him readings which gave Charles a new perspective on life, including the famous Alexander Humboldt's Personal Narrative about a journey to South America. Henslow also persuaded another professor at the college to take Charles as his field assistant to learn how to use equipment and gain some experience studying natural science. When Charles graduated in 1831, he went home and found a letter waiting for him that changed his life forever: An invitation to join the crew of the HMS Beagle.

II

A NEW DOG WITH NEW TRICKS

"A man who dares to waste one hour of time has not discovered the value of life."

— CHARLES DARWIN

When the Beagle left port in 1831, it was only on its second voyage, just like how Charles Darwin was only on his second expedition out into the world of natural science. It was captained again by a name named FitzRoy, who was tasked with surveying the coasts of South America and bringing the data back for future use by the Royal Navy.

Darwin didn't accept the opportunity right away out of deference to his father, who gave numerous reasons why he didn't want his son gallivanting off at sea. However, ever a rational man, Dr. Darwin told his son that if he could find someone whose opinion he respected that thought the voyage was a good idea, he would relent. The young Charles Darwin immediately traveled twenty miles to go hunting with his uncle, Josiah Wedgwood II. When Darwin told his uncle about the voyage, Josiah was thrilled and immediately created counter-arguments to his brother-in-law's objections. Josiah spoke to Dr. Darwin and eventually convinced him to not only support Charles' travels but also to fund them.

❧

However, Dr. Darwin's objections were not the only obstacle Charles needed to overcome. Henslow recommended Darwin to the captain, Robert FitzRoy, who was not looking for a common naturalist but rather an educated gentleman who would keep him company while collecting specimens. To a modern audience, there might not seem like much distinction, but it came down to Captain FitzRoy mainly seeking someone of a good background to accompany him. The position was not so much scientific as social.

❧

The problem was that Captain FitzRoy was a phrenologist. Phrenology is a pseudo medicine that thought the shape and size of the head, face, brain, and cranium-controlled personality characteristics. Upon meeting Charles, Captain FitzRoy decline giving him a position because he believed Charles' nose indicated he was too weak to take on the expedition. It took some time, but Darwin eventually convinced the captain his *"nose had spoken falsely."*

❧

Darwin would not leave for another four months as the Beagle required significant work and refitting, much of which was paid for by Captain FitzRoy. During that time, Darwin stocked his small cabin with scientific books, reading materials, specimen jars, and equipment necessary for collecting samples. After many delays, he set sail with the ship on December 27th, 1831 at 22 years old as one of 73 crewmembers. It was going to be the adventure of a lifetime, but Darwin quickly became seasick and spent a long time in his quarters, regretting his decisions. He would remain ill for a major portion of the next five years, living off a diet of dry biscuits and raisins while on the water.

❧

This journey would last for five years despite originally being planned for two and heavily influence the development of Darwin's theory of evolution. While the Beagle spent its time surveying and charting coasts, Darwin would go ashore and collect samples to study. He investigated the geology and landscapes of different regions in South America and made sure to always work in a private capacity to avoid having to give his collection to someone else. Even from the start, he intended to donate his samples to a scientific institution that would appreciate them. Despite being seasick, he kept copious observational notes about the world around him, primarily about marine biology.

❧

The Beagle's first stop was Madeira Island around January 4th, 1832. The ship intended to make port but was unable to; a squall came in from the west making it too hazardous. Darwin was still sick and nauseous and didn't even notice the storm since he continued to be confined to his quarters. Two days later, the ship tried to dock at the port of Santa Cruz at Tenerife Island but was stopped by a medical ship. The island suffered from an outbreak of cholera, and the Beagle needed to wait through a quarantine of 12 days before coming ashore.

❧

Darwin, again, took little notice until Captain FitzRoy decided they would not wait and would instead continue to the next stop. The decision devastated Darwin, who had made plans to visit a friend and climb atop the island's peak – a dream of his for some time. Captain FitzRoy wrote of the disappointment in his journal but did not reverse the decision out of a desire to keep the voyage moving.

❧

The next stop was the Cape Verde Islands and the island of Santiago. The HMS Beagle anchored at Porta Praya while Darwin and two other crewmembers went on land to the village of Ribeira Grande. Darwin wanted to investigate some Spanish ruins thoroughly and did so but got lost on the return trip to the ship. While on Santiago, Darwin spent his time examining the local cuttlefish and tide pools. His excitement overtook him with the cuttlefish when he discovered they could change colors. He eagerly wrote to his friend, Reverend Henslow, about this intriguing animal. When he returned to England, he learned that this color-changing ability was already known to natural scientists.

❧

Besides cavorting with the cuttlefish, Darwin made his first significant discovery while on the island. He found a white band along a cliff face some 45 ft. above sea level. Upon closer inspection, he realized it was a layer of shells embedded in the rocks yet somehow away from the water. He quickly began to theorize just how something that should have been down by the sea had migrated so far above the waterline.

❧

One of the books Darwin read around this point was Charles Lyell's Principles of Geology, which proposed the truly novel concept of lands slowly rising and falling over immense expanses of time. Lyell developed the idea of uniformitarianism, which suggested that volcanic activity, erosion, and sedimentation occurred at the same rate throughout time, a contrast to the existing idea of a catastrophic start to modern geography.

❧

Lyell's work, along with the geological evidence of shells above the shoreline, would influence a future theory of Darwin's where continents continuously rose while the ocean floor sank. At that moment in 1832 though, Darwin wondered if the land had slowly risen over time since anything more severe or violent would have broken the shell layer.

❧

With these new discoveries, Darwin returned to the ship. The HMS Beagle left Santiago on February 7th and headed towards Brazil, stopping for a day at the known ship hazard of St. Paul's Rocks. Captain FitzRoy wished to examine the exact location of the rocks for the safety of future voyages, so he sent two boats ashore over shark-infested waters to get accurate readings. Darwin went as well to observe the cliffs and rocks and identified two species of bird that live onshore. Apparently, they possessed no predators and were so unfazed by the sailors that it was easy for the crew to walk right up to them and hit them over the head with sticks.

The Beagle crossed the equator on February 16th and arrived in Brazil on February 28th. The tropical rainforests captured Darwin's imagination, and he spent several days enraptured by the lush environment around him. He patrolled the tamer parts of the forests and gushed about the land's beauty in his notes and letters home to his family.

But not everything in Brazil was beautiful to him. Darwin soon quarreled with Captain FitzRoy about the ethics of African slavery and the ethics of treating another human as property. During the period when Darwin grew up, Britain had already outlawed slavery, and it was never popular on the mainland of the country, to begin with. The young man had never actually seen the conditions in which African slaves lived and worked, and Brazil has long been considered one of the most brutal regions.

Captain FitzRoy raged at Darwin and banned him from

sharing dinner with him ever again but reversed his decision after a few weeks. Darwin wrote about the incident, providing an insight into how viewed the entire situation:

> *"We had several quarrels; for when out of temper he [FitzRoy] was utterly unreasonable. For instance, early in the voyage at Bahia in Brazil he defended and praised slavery, which I abominated, and told me that he just visited a great slave-owner, who had called up many of his slaves and asked them whether they were happy and whether they wished to be free, and all answered 'No.' I then asked him, perhaps with a sneer, whether he thought that the answers of slaves in the presence of their master was worth anything. This made him excessively angry."*

❦

The ship's next destination would be Buenos Aires, although it stopped at the port of Montevideo on the way. Sometime during the journey, the ship's surgeon chose to resign because he felt like Darwin usurped his position. During this era, it was common practice for a vessel's surgeon to collect samples, and Darwin had completely taken over. The captain promoted the assistant surgeon to regular surgeon, and Darwin continued to collect specimens for his research and work.

❦

During the journey, a slew of porpoises escorted the Beagle, which Darwin could have enjoyed if he had not been confined to his room from nausea. When they arrived, a postal ship dropped off mail for the ship's crewmembers. Darwin was disappointed neither his family or friends wrote him, but he prepared to ship off his samples for protection and further study.

❧

Around August 17th, he shipped his parcels and notes to Reverend Henslow. Among the collection were rocks, tropical plants, beetles, four different animals preserved in alcohol, and tons of marine specimens all labeled and carefully categorized. Darwin feared the collection would be deemed worthless and fretted for weeks about the value of his work.

❧

After a brief stop at port, Darwin and the crew surveyed the coastline south towards Bahia Blanca, escorted by an Englishman involved in the sealing industry. During a stop at Fort Argentina in Bahia Blanca, Darwin came under the suspicion of the major running the fort. Why? Because he carried a number of strange instruments, the locals never encountered before and the translators could not accurately explain Darwin's job. They described him as a naturalist whose job it was to "***know everything***." The major concluded Darwin was a spy and had him followed by soldiers until he left.

❧

Darwin would make his first major discovery at the Beagle's next stop: Patagonia. Patagonia can best be described as the southern tip of South America, excluding the Falkland Islands but including the Tierra del Fuego. In modern times, it is shared by Chile and Argentina. The geography impressed itself upon Darwin again who, being from England, was accustomed to fields, meadows, cut woodlands, and the city. Patagonia as a region included a diverse array of environments, including deserts, grasslands, lowlands, and sections of the Andes Mountains.

❧

While exploring, Darwin and his assistant Syms Covington found bones and shells encased in a rocky cliff face away from the water. Using pickaxes, the duo and some others dug through the stone and discovered massive fossil bones, including a gargantuan jaw with an intact tooth. Darwin managed to identify the fossil by the tooth; it was a Megatherium, or an ancient ground sloth the size of an elephant.

❧

This discovery was magnificent, but Darwin's powers of observation were what really changed the game and made it so significant. What were the fossils of such an old animal doing next to shells that currently existed but were old enough to have been embedded in the rock? It seems simple enough to a contemporary audience, but to people of Darwin's time, it was unusual. Darwin came to a powerful conclusion: the ancient animal had lived during a time with the same shells as could be found in the 19th century, which pointed to a recent extinction without any climatic or geographical catastrophe.

❧

These early months on the Beagle aptly demonstrate some of

Darwin's strengths, as well as his approach to discoveries. He was a methodical, almost placid man who loved to observe the world he was in and make notes one at a time. When he found evidence to support a new idea, he investigated it further and drew upon new literature and theories without jumping immediately to conclusions, demonstrating an attitude preferable for scientists.

❧

More than anything though, his travails upon the Beagle show just how human the "***Father of Evolution was***." He got seasick and missed adventures. He found himself in fights. He got lost while trying to travel around the islands he visited and frequently found himself escorted back to the ship by the other crewman. He missed his friends and family, and he certainly didn't hit magnificent scientific success right away. It all took time, and he continued to live a regular, if privileged, life.

❧

During the next portion of the Beagle's second voyage, Darwin did not hit upon his grand theory of evolution, even though it created an important baseline for research.

❧

Next, Darwin would go to the famous Tierra del Fuego and the even more well-known Galapagos Islands.

III

THE LAND OF FIRE AND MOCKINGBIRDS

"A moral being is one who is capable of reflecting on his past actions and their motives - of approving of some and disapproving of others."

— CHARLES DARWIN

Tierra del Fuego is the name for the very tip of South America, where numerous islands form a small cluster that comes to a point and merges into the mainland of the continent. The Beagle sighted the area on December 18th. Aboard the Beagle at this time were three natives from the region who had been brought back to England during the ship's first voyage. While in Europe, the trio trained as missionaries and elected to join the crew of the Beagle when it went back to their home.

The Fuegians were ecstatic to be home. When the ship rounded Cape San Diego, dozens of natives approached the shoreline and started to shout at the crew, and the three Fuegians shouted back in their own language. The Beagle sailed along the coast for hours before entering the strait of Le Maire and anchoring at Good Success Bay. In the morning, FitzRoy sent out a party, including Darwin, to speak with a small group of natives willing to come to the shore. They did not appear to speak the same language as the three Fuegian missionaries, so a crude form of communication developed.

The natives and Europeans were cold to one another at first and could not start a dialogue. Finally, some crewmembers presented the Fuegians with bolts of bright red cloth, and the natives warmed up and began to mimic the speech, gestures, and mannerisms of the Europeans. Their ability to copy and repeat entire sentences and even monologues impressed Darwin, but not enough for him to not write unpleasant passages about the people in one of his journals. In particular, he said about the Fuegians:

> *"These poor wretches were stunted in their growth, their hideous faces bedaubed with white paint, their skins filthy and greasy, their hair entangled, their voices discordant, their gestures violent and without dignity. Viewing such men, one can hardly make oneself believe they are fellow-creatures and inhabitants of the same world."*

Here was one of Darwin's biggest character flaws, at least from a contemporary perspective. He protested against slavery and disagreed with the poor treatment of peoples from other places, but he still viewed groups like the natives as being inferior and disgusting in comparison to the Europeans. Ironically, the natives found the European sailors to be the same way, namely because of their pale skin and gross, unkempt beards.

A week after the meeting and spending time prepping the ship and collecting samples, Darwin attempted to hike the terrain. He tried to make his way up a particularly steep hill but became so entangled in the thick vegetation and uneven ground that he became stuck. Eventually, he followed a river back to the Beagle.

The next day, the ship set out again but became mired for weeks in heavy storms and squalls that kept it from landing. Darwin was of little use and spent much of his time nauseous and holed up in his cabin with his work.

The Beagle then circled around the region and Captain FitzRoy was determined to bring the three Fuegians back to their homeland, so they could set up a Christian mission for the other natives. This was no easy task. The Beagle eventually found themselves loading provisions onto whaling boats and taking about 1/3 of the crewmembers out to sea to reach the coast.

❧

Darwin accompanied them and got an even closer look at the lives of the native Fuegians. He still found them contemptible and wondered how they could even be related to Europeans as a species. But that was just it – despite their differences, he still theorized and believed that the natives of South America and the Europeans did have the same origin. His natural imperialist kicked in though as he also thought that the best thing that could be done for the natives was to "***civilize***" them, which meant making them Christian and replacing their culture with that of Europe.

❧

The group set up a mission and locked their provisions in several hunts to try to avoid theft, which was common. They planted a myriad of vegetables and foodstuffs for the mission. Most of the whaleboats returned to the Beagle, while Darwin and some of the crewmembers, including Captain FitzRoy, took another to finish some surveying in the area.

❧

When they returned to the mission, they were horrified to discover that the native Fuegians had completely demolished and ransacked much of the mission, stealing clothes, food, and tools. FitzRoy gave up and left the three Fuegian missionaries to fend for themselves. Five days later, FitzRoy guiltily went back to check on the missionaries and was delighted to see that the building was cleaned up, the garden replanted, and the overall area back in business. The Beagle stayed in the area for a few more days before disappearing and heading for a new location: the Falkland Islands.

❦

The ship arrived around March 1st and found itself assisting the British navy, who had taken over the region from Argentina just a few weeks ago. They waited until reinforcements arrived, during which time Darwin went onshore and found some more intriguing tidbits for his growing collection of notes. Namely, he noticed that the fossils found on the islands were very different from those on the mainland. A small observation that made him decide to compare his samples more closely. This event would influence his views on plant and animal distribution and contribute to his idea that the same animal would adapt differently to new environments.

❦

The Beagle would spend a long-time surveying South America and returned to it several times during the ship's five-year voyage. Darwin grew more and more into the role of an adventure as he began to traverse more terrain and interact with the people living throughout the Tierra del Fuego and the Galapagos. He rode with gauchos, collected specimens, survived an earthquake in Chile, and climbed mountains. He still struggled to fill such a role though, as could be seen in how he tried to learn how to throw a bolo-like the gauchos and wound up capturing his own horse.

❦

Throughout this time, seashells were his constant companion. He found strips of them in the mountains and in cliff faces wherever he went. He noticed that mussel bands could be found in fertile hills and that fossilized trees were on sandy, barren beaches. The environment did not match the world 19th century Europeans had created, and each observation began to weigh more and more heavily upon Darwin and his ideas.

❦

Further, he found the same kinds of animals in South America that he did in Europe. Many people at the time believed God had created animals specifically suited to each environment. Darwin wondered how two such different environments could produce the same creatures with such slight differences if this was the case.

❦

The HMS Beagle surveyed the Galapagos Archipelago from September 15th to October 20th, 1834. During this period, Darwin was able to hop from island to island, collecting samples and examining the different creatures which lived in the area. The finches, which would be named after him, mockingbirds, and the turtles pressed upon him the most.

❦

All three creatures could be found on the varying islands, but all with the differences between them. The mockingbirds came in different colors and sang separate songs to attract one another while still being the same basic creature. The sea turtles possessed slight variations in the shape and design of their shells which made it easy to tell which island they originated from. Despite popular belief, Darwin did not actually pay attention to the finches and would only identify differences in them when samples were brought back to England by an ornithologist unrelated to the Beagle expedition.

❦

It seemed strange to Darwin that animals which lived in such close proximity to one another and bore so many similarities could still have basic differences. The idea started to percolate in his mind, although he would not understand the significance of his discovery until 18 months after the expedition ended.

❦

The second voyage of the Beagle continued for several more years and traveled around the southern half of the globe, making stops throughout South America and in Australia and Africa. Darwin collected samples the entire way and continued to write down notes about the animals he encountered, including how striking it was that somehow kangaroos, and platypuses existed on the same planet.

❦

Throughout the journey, Captain FitzRoy's surveying of coastlines continued to support Darwin's nascent hypotheses about the natural world. By the end of the trip, Darwin recorded in his notes that his observations seemed to upset the very nature of the world as Europeans knew it. If animals could have the same basic ancestor and then change over time to suit their environments, it meant that no grand creator had set out making the perfect animals for different climates. It would, as Darwin wrote:

“undermine the stability of Species.”

IV

DARWIN GOES TO LONDON

"The very essence of instinct is that it's followed independently of reason."

— CHARLES DARWIN

When the Beagle reached Falmouth, Cornwall, Darwin was a celebrity among scientists. The ship docked on October 2nd, 1836. Darwin's first stop was home to Shrewsbury, where he spent time with his family and was delighted to discover he had won the approval of his father. Darwin chose not to become a clergyman and instead pursue a growing scientific career that had been fostered by Reverend Henslow while Darwin was at sea.

During the voyage of the Beagle, Henslow took the specimens and notes Darwin sent him and started showing them to select groups of interested scientists. He even presented the materials to the Cambridge Philosophical Society. The other scientists admired Darwin's work and wished to speak to him about his discoveries. After some time in Shrewsbury, Darwin left for London.

Henslow encouraged Darwin to get out and introduce himself to the scientists and naturalists interested in his research. Upon arriving in London, Darwin was surprised to hear that the giant sloth skeleton he found in South America was displayed at the 1833 meeting of the British Association of the Advancement of Science. His first major task in London was finding naturalists who would be willing to catalog and describe the over 4,000 specimens he sent back to Britain over the last five years – no easy feat since most scientific research was backlogged.

Of course, being a gentleman scientist did not produce any money. Darwin relied on his father to gather donations from others and support his research, so he didn't have to work another job in the interim. Much of this money went to pay the naturalists he found. He didn't catalog the specimens himself due to the sheer amount of time it would have taken, as well as the negative effects it could have on his career.

If he spent all his time at a university sorting through

samples, then he wouldn't be able to make the social connections necessary to advance, and Darwin wanted to advance. For most of his young life, he drifted between passions and jobs, unable to really find what he wanted to do. Now that he experienced success, gained the respect of others, and was making legitimate scientific progress, he wanted to continue.

❧

While in London, Darwin met Charles Lyell, the man whose books he read while at sea. Lyell served as the president of the Geological Society and introduced Darwin to other famous people like Richard Owen, the comparative anatomist. Through his connections, Owen convinced the Royal College of Surgeons to work on the fossils Darwin brought back. Darwin also gained membership in the Geological Society and the Athenaeum Club.

❧

Darwin headed to Cambridge and stayed there to work. Unlike his mannerisms when he was young, the naturalist was quickly becoming a workaholic who devoted most of his time to his studies. He prepared a presentation for the Geological Society that suggested the landmass of Chile was rising. He read it with Lyell's backing and additionally presented his animal specimens. An ornithologist pointed out that the '***strange birds***' Darwin brought back – and which he thought were separate types of bird-like blackbirds and finches – were all actually finches. He was the one who inspired Darwin to look more closely at the species of finches for which Darwin would become famous.

❧

Darwin returned to London in March. He joined Lyell's group of friends and scientists while also taking time to hang out with his progressive, free-thinking brother. He lodged in Great Marlborough Street and made many new acquaintances and compatriots who supported his scientific ideas. Among them was Harriet Martineau, a social reformer, author, and feminist who was also a Unitarian that supported the idea of the transmutation of species. The concept of the transmutation of species preceded Darwin's work that said species came about from spontaneous generation (not a common ancestor) and then developed characteristics to suit their environments.

Around July, Darwin began to fill journals and notebooks with his ideas about what the transmutation of species could be – the term evolution was not used at this time. In one, he sketched a rough tree with branches coming from it that would form the basis of his future evolutionary tree. Above were written the words "***I think***." His main speculations were that one species could change into one another and needed to adapt to the environment. He additionally dismissed the ideas of Lamarck, the creator of the original transmutation of species, who had claimed that lifeforms evolved to be higher than one another.

During this time, Darwin additionally edited the journals he kept while sailing on the Beagle. Captain FitzRoy wished to publish his survey data and thought Darwin's observations should be included as well. Originally planned to be part of a two-volume work, Darwin's notes eventually became their own book in the set. He also started editing all the reports on his collections and acquired funding from the Treasury for his work. To add to his workload, Darwin also made ridiculously optimistic agreements with publishers about when he could have his works on geology finished.

❧

On September 20th, 1837, Charles Darwin suffered from a heart palpitation that his doctor attributed to his overwork and stress. Multiple doctors encouraged him to go to the country for a few weeks of rest and relaxation, so he chose to visit Shrewsbury for a short time before visiting his Wedgwood cousins at Maer Hall in Staffordshire. He struggled to get comfortable because his cousins and relatives wanted to hear about his adventures on the HMS Beagle.

❧

He did, however, meet his cousin Emma Wedgwood once again. She served as a nurse to Darwin's aunt and impressed him with her poise, grace, and quick mind. Emma was technically Charles' first cousin, which seems odd to modern audiences but was very common for wealthy British. By 1837, Emma had already turned down several marriage proposals from different men and was difficult to impress.

❧

By this point, Darwin, as a young man nearing 30, had considered marriage and written lengthy notes about whether it was a good idea. Perhaps embarrassing for him, but modern researchers do have access to his musings. Among his scribblings about marriage was a list about whether it was a good idea, both for his health and career:

❧

"This is the Question

Marry

Children (if it Please God) Constant companion, (& friend in old age) who will feel interested in one, object to be beloved & played with. Better than a dog anyhow. Home, & someone to take care of the house. Charms of music & female chit-chat. These things are good for one's health but terrible loss of time.

My God, it is intolerable to think of spending one's whole life, like a neuter bee, working, working, & nothing after all. No, no won't do. Imagine living all one's day solitarily in smoky, dirty London House. Only picture to yourself a nice soft wife on a sofa with good fire, & books & music perhaps. Compare this vision with the dingy reality of Grt. Marlbro' St.

Marry—Mary—Marry Q.E.D.,

Not Marry

Freedom to go where one liked. Choice of Society & little of it. A conversation of clever men at clubs. Not forced to visit relatives, & to bend in every trifle. To have the expense & anxiety of children —perhaps quarreling— Loss of time. cannot read in the Evenings—fatness & idleness— Anxiety & Responsibility. Less money for books & if many children forced to gain one's bread. (But then it is very bad for one's health to work too much)

Perhaps my wife won't like London; then the sentence is banishment & degradation into indolent, idle fool.

It being proved necessary to Marry.

When? Soon or Late?

The Governor says soon for otherwise bad if one has children— one's character is more flexible—one's feelings more lively & if one does not marry soon, one misses so much good pure happiness.

But then if I married tomorrow: there would be an infinity of trouble & expense in getting & furnishing a house, —fighting about no Society— morning calls—awkwardness—loss of time every day. (without one's wife was an angel, & made one keep industrious). Then how should I manage all my business if I were obliged to go every day walking with my wife. — Eheu!! I never should know French,—or see the Continent—or go to America, or go up in a Balloon, or take solitary trip in Wales—poor slave.—you will be worse than a negro— And[24] then horrid poverty, (without one's wife was better than an angel & had money)— Never mind my boy— Cheer up— One cannot live this solitary life, with groggy old age, friendless & cold, & childless staring one in ones face, already beginning to wrinkle.— Never mind, trust to chance—keep a sharp lookout— There is many a happy slave"

Yes, Darwin had a lot of thoughts on marriage. His overall consensus, though, seemed to be that marriage would be a good idea, if only for his own happiness and peace of mind.

He set his sights on Emma but feared their different views about life could turn the marriage sour. She was staunchly religious, especially after the death of her sister. She strived to see her again in Heaven and feared being separated from her husband forever if he was not Christian. Darwin and Emma eventually came to a compromise where he would at least have faith in Jesus, and they would avoid discussing the topic too much.

❧

Darwin and Emma married on January 29th, 1839 at St. Peter's Church in Maer. They received plenty of money from their family to start a wealthy life together. Less than a year later, their first son was born. Although Emma disagreed with some of Darwin's scientific research, she still read and edited his manuscripts before they were submitted to scientific societies and publications. Unfortunately, around this time was when Darwin's health started to worsen. He frequently suffered from vomiting, stomach pains, boils, headaches, and heart palpitations. For the rest of his life, his dear Emma would play the role of the nurse to try to keep him healthy.

V

THE EVOLUTION AND TRAUMA OF THE "THEORY OF EVOLUTION"

"To kill an error is as good a service as, and sometimes even better than, the establishing of a new truth or fact."

— CHARLES DARWIN

Before his marriage, Darwin had already started to solidify his theory of natural selection, where organisms with the most suitable traits would be most likely to survive in any given environment. After getting married, he focused almost entirely on the concept of natural selection and began to experiment and study the selective breeding of plants and animals. Again, he was ecstatic to discover that most species were not static and were, in fact, subject to variation.

❧

Around May 1839, Captain FitzRoy's accounts of the second voyage of the ***Beagle*** were published and gained instant acclaim. Darwin's section, ***Journal and Remarks*** became so successful that it received its own publication and subsequent rerelease a couple of years later. Darwin wrote to Lyell when his notes were republished and pointed out that while FitzRoy was his friend, the man never did seem to agree with him on anything, including his observations about the potential origins of species.

❧

Darwin next published a book entitled ***The Structure and Distribution of Coral Reefs*** in 1842. This volume covered his theory of atoll formation and took three years to write. It was a moderate success and emboldened Darwin to continue his work on the theory of natural selection. It was not an easy road, especially since he feared the criticism and scorn he might receive from other professionals.

❧

His family moved out of London for separate reasons and took up in the rural Down House where they could escape the pressure of the city. Darwin aspired to visit London a couple of days each month to avoid becoming a country mouse but found that the carriage ride to the train station made him ill. Around this time, Darwin commented that trying to explain his theory about natural selection to his friends was like confessing to murder. He felt guilty and almost sinful going against the standard Victorian narrative.

However, many other people were supportive, including Darwin's friend, a botanist named Joseph Dalton Hooker.

By July of 1844, Darwin had over 200 pages of what he deemed his "***sketch***" of natural selection and instructions about how it should be expanded if he died early. His fear of an early death grew worse and worse with each passing year due to his chronic illness, which just seemed to grow steadily more severe over time. Still, the natural scientist kept going. He kept up numerous correspondences with his colleagues and friends back in London, writing over 7,000 letters and receiving that many in return.

Although a busy man, Darwin kept a simple, easy schedule while in the countryside. He woke up early to take long walks before breakfast and ate his first meal around 8:00. Darwin then worked for roughly one and a half to two hours in his private study before listening to Emma read family letters. He then disappeared to work and take a stroll with Polly, his fox terrier, and companion. He returned for lunch around 1 o'clock and then read, studied, and wrote letters until 3:00 when he would take a break and listen to Emma read. More work followed, then dinner, and finally a full session of relaxation in the evening where he played backgammon with Emma or listened to her play the piano.

It's clear from Darwin's schedule and his regular correspondences that he loved his wife dearly and spent a good portion of his day with her. He also showed an affectionate competitive streak, keeping track of their backgammon victories and gleefully writing to a friend that Emma only won 2,490 games compared to his 2,795. He also loved his children and was the exact opposite of the distant Victorian father archetype. He made time in his day to play with them and never shouted or scolded when they entered his office or bothered him when sick.

Still, Darwin remained preoccupied with his work. Although he showed it to the eminent Hooker, he feared allowing anyone else to read it before it was finished. This terror became exacerbated after the anonymous publication of a book called ***Vestiges of the Natural History of Creation***. The book caused a massive public uproar for suggesting there was a perpetual transformation of species without a grand creator. Although Darwin enjoyed the idea, he panned it along with his fellow professionals for having major geological and zoological errors.

Darwin, therefore, published several more books while working on his pet project, including one in 1846 about geology. He also diversified his hobbies, becoming once more intrigued with the marine invertebrates which occupied his school years. Around 1847, Darwin sent his notes off to Hooker for review. Hooker offered constructive criticism and critique but continued to question whether Darwin's theories held any weight against creationism.

Darwin suffered considerably over the next two years, drastically affecting his work. First, his father died in 1848, which devastated Charles. Then, his illness grew to the point where he spent days vomiting and unable to move. To try to combat the severity of his symptoms, Darwin went to a Malvern spa for a treatment. While there, he found that hydrotherapy helped considerably and started to wash his head with cold water in the mornings to simulate the treatments at the spa. Unfortunately, just as Darwin was feeling better, a new blow would strike in his life.

❧

By this point, Darwin had several children with Emma. His second child and oldest daughter, Annie, became chronically ill with symptoms similar to the ones he suffered. He feared that it was his fault and that the malady he suffered was hereditary, although many modern scholars attribute Annie's failing health to scarlet fever and then tuberculosis. Desperate, Darwin escorted his ten-year-old daughter to the same spa he attended for hydrotherapy. Annie died along the road.

❧

Annie had always been the apple of her father's eye. He wrote fondly of her and frequently told his friends how much he loved her spontaneous affection and sweet tidiness, or her habit of making clothes and ribbons for her dolls and imaginary worlds. He wrote in one of his memoirs:

> *"We have lost the joy of the household, and the solace of our old age... Oh that she could now know how deeply, how tenderly we do still & and shall ever love her dear joyous face."*

The natural scientist's grief was only slightly lessened by the birth of another son, Horace, less than three weeks after Annie's death. Darwin slowly resumed his research about barnacles to contribute to his theory of natural selection. The idea was that if he could give concrete evidence for one subgroup of organisms, it would be easier to convince people of a universal theory.

After eight years of barnacles, after which Darwin commented that he hated them more than any other man in the world, he seemed to have enough evidence for his ideas. He discovered numerous homologies that showed how changed body parts fulfilled different functions for new environmental conditions. He further found an intermediary stage between the male and female sex, which indicated there was some evolution between the two.

Darwin's contributions earned him the Royal Society's Royal Medal and praise from his colleagues. It truly made his reputation as a biologist, more so than his voyage on the ***Beagle***. He became a fellow of the Linnean Society of London and could then access their library through the mail, which gave him an even greater pool of resources to draw from. He started to reassess his theory about species and decided that the divergence between creatures could be explained by a need to adapt to new environments.

VI
ON THE ORIGIN OF SPECIES

"I am turned into a sort of machine for observing facts and grinding out conclusions."

— CHARLES DARWIN

Around 1856, Darwin moved away from his work with barnacles and started to experiment with plants, germination, and whether eggs or seeds could withstand travel across seawater. At this point, he and his family were financially stable and secure, having inherited an enormous sum of money from his father's estate and invested in railroads and canals. Darwin's research proved successfully when he found that some seeds could still germinate even after being submerged in water for over a month. He published his findings in the ***Gardener's Chronicle***.

❧

Darwin's friend Hooker became more and more convinced in Darwin's ideas about the potential evolution of species, and several others began to question whether Darwin would extend his ideas to humans. Darwin hesitated. He found the subject to be too fraught with prejudices and potential dangers to want to include it, although he did layout sections tentatively in his draft about human evolution and the development of different races.

❧

He also faced some external pressure. A man named Alfred Russel Wallace published a paper entitled "***On the Law which has Regulated the Introduction of New Species***." It was not exactly the same as Darwin's work, but Lyell saw some eerie similarities and encouraged Darwin to please publish something, so he could have precedence in the field. Ever a plodder, Darwin started a paper but found himself stopping because of difficulties in simplifying the material and answering tough questions.

❧

On June 18th, 1858, Darwin's inability to publish his work came back to bite him when Wallace, who was a friend of Darwin's and in contact with him despite doing fieldwork in Borneo, sent a manuscript through the post. This paper described natural selection, and Wallace requested that Darwin send the document on to Lyell, their mutual colleague. Shocked and dismayed, Darwin still did as Wallace requested and even offered to have it published in a source of Wallace's choosing. Wallace, currently facing disaster as the inhabitants of Borneo suffered from an outbreak of scarlet fever, left it in his friend's hands.

❧

Together, Lyell and Hooker decided on a joint presentation of the work of Wallace and Darwin at the Linnean Society on July 1st. The presentation would be called ***On the Tendency of Species to form Varieties and on the Perpetuation of Varieties and Species by Natural Means of Selection***. Darwin did not attend for, on June 28th, he lost another child to scarlet fever and was too distraught to go. When Wallace found out that his scientific colleagues published Darwin's work alongside his, he took it with grace and recognized that Darwin spent many more years on the same subject.

❧

Despite not going to the presentation, Darwin eagerly awaited reviews of his and Wallace's work. Unfortunately, the other scientists did not seem impressed. The president of the Linnean Society believed there had been no revolutionary discoveries or presentations in 1858, while others thought the subjects presented had been done to death. Rankled, Darwin worked even harder to create his magnum opus: ***On the Origin of Species.***

❧

It took 13 months and numerous bouts of severely declining health for the book to be finished, but Darwin received constant feedback and encouragement from his friends. Lyell even arranged for the book to be published by John Murray – a British company and not, in fact, a person. ***On the Origin***

of Species was popular from the outset, selling all the original 1,250 copies. The book didn't actually use the term '***evolution***' and avoided any discussion about the possibility that humans could be included in Darwin's theory, but he did add a cryptic line at the end about shedding some light on the origins of humans.

The book spawned international interest and surprisingly less controversy than ***Vestiges of the Natural History of Creation***. Unlike the other book, Darwin's ***On the Origin of Species*** included more research and evidence to support its claims, which earned it more support from the scientific community. Darwin was too ill to attend the public debates about his book, but he did keep newspaper clippings, reviews, articles, caricatures, satires, and the comics made about it. He also ramped up his correspondence with other professionals, eager to know what everyone thought.

Although the book avoided discussing human origins, it had too many lines or side notes from which people could infer that humans might be included in the theory. Indeed, many reviews of ***On the Origin of Species*** went with this point. Some early negative readers wondered what it would mean for humans if they had evolved from monkeys and decided that such questions should be left to theologians, not regular people.

Thomas Huxley, an English biologist who would earn the

nickname "***Darwin's Bulldog***," was among the first to submit a positive review. In doing so, he attacked the work of other well-known members of the field, including Richard Owen, his long-standing enemy. Owen was one of the best analysts of fossils and didn't disagree with Darwin but did believe evolution was far more complex than Darwin's theory. Huxley's swipes did not go unnoticed, and Owen publicly dismissed Darwin's work and started to work with others on the idea of a supernaturally guided evolution and natural selection instead.

Owen angered Darwin, not because he disagreed with Darwin's theories but because Owen chose to attack Darwin's friends and was downright condescending in his review. Ever a patient man, Darwin could handle criticism but disliked personal attacks and arrogance from his fellow scientists. To make matters worse, Owen had been a friend of Darwin's who seemed to react more out of spite than any actual scientific reasoning, an action which ended the two men's friendship and association.

On the Origin of Species received mixed views from the Church of England. Darwin's former Cambridge tutors and friends, including Reverend Henslow, dismissed his ideas because they were incompatible with their religious beliefs. However, his theory did gain ground with the younger, more liberal clergymen emerging during the 1850s. They interpreted natural selection as just another part of God's grand design for the world and something completely compatible with an almighty deity. In 1860, the publication of a series of essays written by liberal Anglican theologians about the historical criticism of religious documents distracted the church from Darwin.

Scientific confrontation and debate continued for years after Darwin's publication. Queries about whether the supporters of Darwin's theory really wanted their relatives to be considered apes were thrown about in public and numerous satirical cartoons entered newspapers and journals. Most people were unable to embrace Darwin's ideas to the fullest, including Darwin's longtime friends like Hooker, Huxley, and Lyell. Still, they continued to support and champion his ideas, believing they possessed a new legitimacy all their own.

A new movement evolved called Darwinism, which supported evolution in its many forms. Lyell and Huxley published books with evidence to support Darwinism, including correlations between the skeletons of humans and apes. Lobbying by these men and others earned Darwin the prestigious 1864 Royal Society's Copley Medal. On the same day, it was given to Darwin, Huxley arranged a meeting of a new scientific society called the X Club, which was about progress and discovery without religious constraint. By the end of the 1860s, the majority of scientists agreed that some form of evolution existed, but many disagreed with the concept of natural selection.

Darwin stayed out of the public eye from 1862 until 1866

because of his illnesses but tried his best to remain abreast of all the new developments his work caused. During this time, he grew his famous full beard. Cartoonists took this development and ran with it, drawing caricatures that depicted Darwin as a hairy half-human/half-ape. While at home, he continued his hydrotherapy treatment and research, this time focusing on pigeon breeding. Ever methodical, he remained undeterred by criticism and continued to find more evidence for his theory.

***

On the Origin of Species was not only a success in Great Britain but around the world. It received translation into 11 separate European languages during Darwin's lifetime, and the publishing company made versions in Japanese and Chinese by the end of the 19th century. Darwinism became a fixture of popular culture and resonated with many of the budding scientific and social movements of the time because it gave humans more freedom and control.

VII
THE DESCENT OF A MAN

"A scientific man ought to have no wishes, no affections, - a mere heart of stone."

— CHARLES DARWIN

For the next 22 years, Darwin suffered from his illnesses in silence and attempted to continue work on what he deemed to be his "***big book***." He considered ***On the Origin of Species*** to be an abstract of what he was truly capable of and dove further into research. This time, he focused on worms and plants, often incorporating his sons and Emma into the projects. He published edits of his previous works and many new papers and writings, including ***Descent of Man, and Selection in Relation to the Sexes*** in 1871.

❧

In ***Descent of Man***, Darwin set out to provide as much evidence as possible that humans are animals and that impractical or inconvenient traits are the result of sexual selection. He further added elements like the ideas of physical and cultural racial classification and inherent differences between the sexes of male and female. Although he began to apply his theory to people, he stressed that all humans were the same species – partially to avoid the inevitable social fallout and consequences that would come from applying the science of evolution to a racist society.

❧

The ***Descent of Man*** was the first official document that used the term "***evolution***" to describe Darwin's theories. He followed up the work with another one in 1872 called ***The Expression of the Emotions in Man and Animals***. This book explained the evolution of human psychology and how it mirrored animal behavior.

❧

Although pleased by the response to these publications, Darwin was troubled with how people started to take his ideas and apply them to societal issues. For example, he did not view native peoples to be another species, but he had no problem with imperialism and did little to defend the races and peoples who would be targeted by programs and actions influenced by his work. Some regard this aspect of his character as being someone who throws a match at a building expressed sadness at it being on fire and then walked away.

❧

Darwin was further surprised by how well both volumes were received and by how little criticism he received in response. It seemed like many individuals in the scientific community gladly embraced Darwinism. He received an honorary doctorate from Cambridge University in 1877 and numerous other awards and honors, but he remained too controversial to be knighted by Queen Victoria for his work.

❧

Even with his failing health, Darwin's death happened suddenly. He recorded that on Christmas Day in 1881, he suffered from severe chest pains that left him barely able to walk and move around. Although he attempted to regain his health over the next three months, the condition only worsened. Doctors from all over the country came to treat him, including the royal physician, Dr. Andrew Clark. Clark refused payment, stating it was an honor to treat someone so influential and famous.

❧

On April 18th, 1882, Darwin suffered from a massive heart attack while in bed. He lost consciousness and took a long time to be revived. While awake, Darwin seemed to sense his own death was approaching and stated that he was unafraid since it was only natural. He died at 4:00 pm the next day at 73 years old, leaving behind Emma and his seven living children.

❧

He and Emma both wanted him to be buried next to the little church in Downe where Down House was located, but Darwin's scientific colleagues and many members of Parliament insisted that he be laid to rest in Westminster Abbey after a grand ceremony. The pallbearers of the coffin were many of Darwin's old friends, including Wallace, Hooker, and Huxley. Even one of the American ambassadors assisted. People from all around the Western world showed up, read speeches, and reminisced about a scientist they never met. Emma, in her old age, could not attend her husband's funeral.

After Darwin's death, many apocryphal stories popped up about his own views about the theories he created. Many of his enemies started to spread rumors that he recanted his own theory upon his deathbed or could be seen reading the Bible and saying hymns – despite becoming less and less religious with each of his scientific discoveries. Against these stories, Darwin's friends and colleagues continued to champion his theories, and his notes spread and became more and more popular and well-regarded.

Darwin's children published his autobiography, memoirs, and the letters he wrote to them after his death to celebrate his legacy and give the public a better view of the man who changed the world. They were also his staunchest reporters, frequently countering the rumors about their renouncing his ideas. They additionally ensured his remaining notes were published or went to other scientists. Emma, distraught by her husband's death, also strove to maintain Charles' legacy.

What a legacy it is.

VIII

AN UNDESERVED LEGACY?

Like any human, Charles Darwin possessed unique strengths, weaknesses, virtues, and flaws. In the contemporary 21st century, he is regularly seen as a genius and is frequently listed as the most influential biologist of all time, and with good reason. His theory of evolution laid the groundwork for the experiments of other scientists interested in biology and the natural world. Aside from that, he also contributed to fields like geology and founded the study of the expression of emotions.

Among his modern accolades and awards include:

- A section of water that adjoins the Beagle Channel of the Tierra del Fuego Archipelago was named the Darwin Sound after he acted quickly and saved several crewmembers of the *HMS Beagle*

- A Mount Darwin in the Andes named for his 25th birthday
- The naming of Port Darwin in Australia in 1839
- Another Darwin Sound was named in 1878 in the Queen Charlotte Islands
- The renaming of the city of Palmerston in Port Darwin to Darwin in 1911
- Over 250 species of creatures named after Darwin
- Nine genera (the plural of scientific genus) named after Darwin
- A complete fossil primate over 47 million years old was named *Darwinius* in 2009 because it shows the transition between species
- An actual day called Darwin Day to celebrate his birthday and accomplishments
- The creation and renaming of numerous institutions, including Charles Darwin University, Charles Darwin National Park, the Charles Darwin Foundation, and the Charles Darwin Research Station
- The creation of two separate awards, one to celebrate major advances in evolutionary biology and one jokingly awarded to individuals deemed too stupid to be in the gene pool

In addition to all of these are numerous streets named after him, statues, special commemorative issues of coins, academic conferences, festivals, and the regular academic citations and credit given by other researchers and scientists. He also has dozens of books written about him – including this one – and at least two movies.

STRENGTHS AND VIRTUES

As far as his career and accomplishments are concerned, Darwin's greatest strengths were his observational and critical thinking skills. He was arguably not the most creative thinker, but he was able to take the phenomena he saw in the natural world, sort through it, and find patterns. His arguments were brilliant and even went over the heads of other eminent scientists of the time because they were so unique and innovative. No one had thought of them before, and that truly put Darwin above many others during the 19th century.

Perhaps, Darwin's greatest virtue was his overwhelming modesty and patience, despite everything that happened following the publication of his work. Even while being lambasted by critics and even former friends, Darwin was calm. He did not attack his critics and actually encouraged many of them, although he made an exception for Owen. He lived to hear the reviews of other people in scientific fields and even adjusted his ideas based on the evidence and work of others.

❧

His letters to his family, published posthumously by his children because they wanted the public to get a better idea of their father, showed a similar character. He was consistently one to downplay his achievements and attribute them more to luck and his attention to detail than anything else. Indeed, he comes across as a man who was truly surprised by his accomplishments and success and didn't want to be some unusual genius. Admittedly this could be attributed to the Victorian tendency to avoid looking vain about one's accomplishments, but Darwin's friends and family state he was always this way.

❧

Darwin was additionally progressive during his time. He opposed slavery and refused to allow his theories to be applied to the races. His friendship with the taxidermist John Edmonstone convinced him that black people could have the same mental faculties as whites, and he extended this thought to other people as well. This seems like basic sense and decency to a modern audience but was unusual for a British man of the 19th century.

❧

One last virtue: He was a faithful man, both to his ideals and to his wife. While he was always willing to listen to new ideas that supported or weakened his theories, he stuck with what he thought was true – so long as it was supported by evidence. Not being affected by the uneducated or society, but still listening to other professionals, laid much of the groundwork for Darwin to be an excellent scientist. Such habits fit with the character he demonstrated in his youth, which was one of almost plodding and painstaking observation. He was not a dreamer. He was an observer.

Outside of his career, he was also extremely faithful to his wife, Emma Wedgwood, and his children. It's significant to point this out since so many famous individuals – men, women, politicians, actors, musicians, poets, authors, athletes, etc. – possessed less than admirable romantic or parental sensibilities. Despite some initial hesitation to marry, Darwin did eventually propose and remained with his sweetheart until his death, having children and striving to give them every advantage in life.

WEAKNESSES AND FLAWS

Sometimes it can be difficult to truly determine the flaws of a man like Charles Darwin because he was a controversial figure during his lifetime. Many viewed him as a raving lunatic or heretic, and some quickly lambasted him. The easiest way to discover his weaknesses is by looking at his personal correspondences, notes, and recorded conversations with others. Based on these, there actually is a new train of thought in the academic world, which states that Darwin's greatest weakness was that he was a plodder.

But what does this mean?

As a plodder, the idea is that Darwin, despite such great

success in the field of evolutionary biology, was not creative and worked slowly. His ideas came from constant observation, not any strokes of genius or unusual thinking. This doesn't seem like a weakness, but modern audiences do view it as a flaw because if he managed to create something like the theory of evolution through observation, what could he have done with a little more intuitiveness? However, this shouldn't be focused on too much since it's more of an "***academic***" detail and doesn't affect his overall character or work.

In a more general sense, Darwin almost lost out on several publications because he spent so much time trying to perfect his ideas. A good chunk of his delays could be contributed to his poor physical health, but he was also an indecisive person and struggled to publish something if he didn't think it was perfect. Although individuals should take pride in their work, check it for errors repeatedly, and strive to be thorough, there comes the point where this can be a major block in the creative, professional, and scientific world.

Some flaws Darwin definitely did have directly related to his upbringing – like most wealthy white men in Great Britain, he was classist, sexist, and an imperialist. In particular, he was not concerned about the dire working conditions of Britain's lower classes and attributed it to another form of natural selection. While he vehemently opposed slavery, he viewed colonization and the death of native peoples as inevitable due to the theory of evolution. While saddened, he did nothing about it because he supported British expansion at the expense of others. Finally, he believed men were naturally superior to women, a common thought at the time, but one to which he again applied his theory of evolution.

❧

Something a lot of people also don't know is that Darwin's half-cousin was Francis Galton, a name anyone familiar with the social history of Europe, the United States, and Latin America should know. Galton was intrigued by the idea of applying Darwin's theories to humanity and suggested that humans could be bred to encourage positive physical and mental traits. Darwin supported the idea in theory, but not in practice because he believed it would stymy the best in humanity. Galton would take these ideas and create eugenics.

❧

The incidents with Dalton, class relations, gender differences, and imperialism all point to perhaps Darwin's greatest flaw or weakness – he was passive. He supported some progressive ideas and in general seemed to hope his fellow humans would do the best, but he never seemed to do anything about it. He put forth the theory of evolution and protested its manipulation in some areas, but not in others, which led to some odd conclusions based on morality and general wishy-washiness. One the one hand, he seemed to want nature to take its course and continue its evolution. On the other hand, he didn't approve of it when it didn't suit his needs.

IX
CONCLUSION

So, Darwin's actual weakness?

He was very much a human influenced by his time, perhaps more progressive and liberal than others. It's hard to judge him through a modern lens, especially regarding his views on race, gender, and colonization. What someone could take away from this is to try to be a consistent individual and stick to their convictions.

In terms of using Darwin's strengths, people should definitely follow his pattern of ignoring criticism from people outside of their field but listening to individuals who do have some form of knowledge or experience relative to the situation. For example, someone who visits a doctor should listen to that opinion, but not one from someone on the internet. A scientist or researcher should take advice from other professionals and not from strangers.

❧

His persistence is also admirable and can help individuals stick to their guns and continue their work. The fact that he was willing to stop if proven wrong listened to his teachers and pursued his dreams are also great practices for a regular person.

X
FURTHER READING

- David Quammen's The Reluctant Mr. Darwin: An Intimate Portrait of Charles Darwin and the Making of His Theory of Evolution.
- Tim M. Berra's Charles Darwin: The Concise Story of an Extraordinary Man.
- Charles Darwin's The Autobiography of Charles Darwin.

YOUR FREE EBOOK!

As a way of saying thank you for reading our book, we're offering you a free copy of the below eBook.

Happy Reading!

GO WWW.THEHISTORYHOUR.COM/CLEO/

Made in the USA
Monee, IL
19 July 2023